For Brian G.
M. J.

First U.S. edition 2019

Library of Congress Catalog Card Number pending
ISBN 978-0-7636-7538-7

18 19 20 21 22 23 LEO 10 9 8 7 6 5 4 3 2 1

Printed in Heshan, Guangdong, China

This book was typeset in Avenir.
The illustrations were done in mixed media.

Candlewick Press
99 Dover Street
Somerville, Massachusetts 02144

visit us at www.candlewick.com

BEWARE OF THE CROCODILE

illustrated by

MARTIN JENKINS   **SATOSHI KITAMURA**

CANDLEWICK PRESS

If there's one thing you should know about crocodiles,

it's that they're really scary—

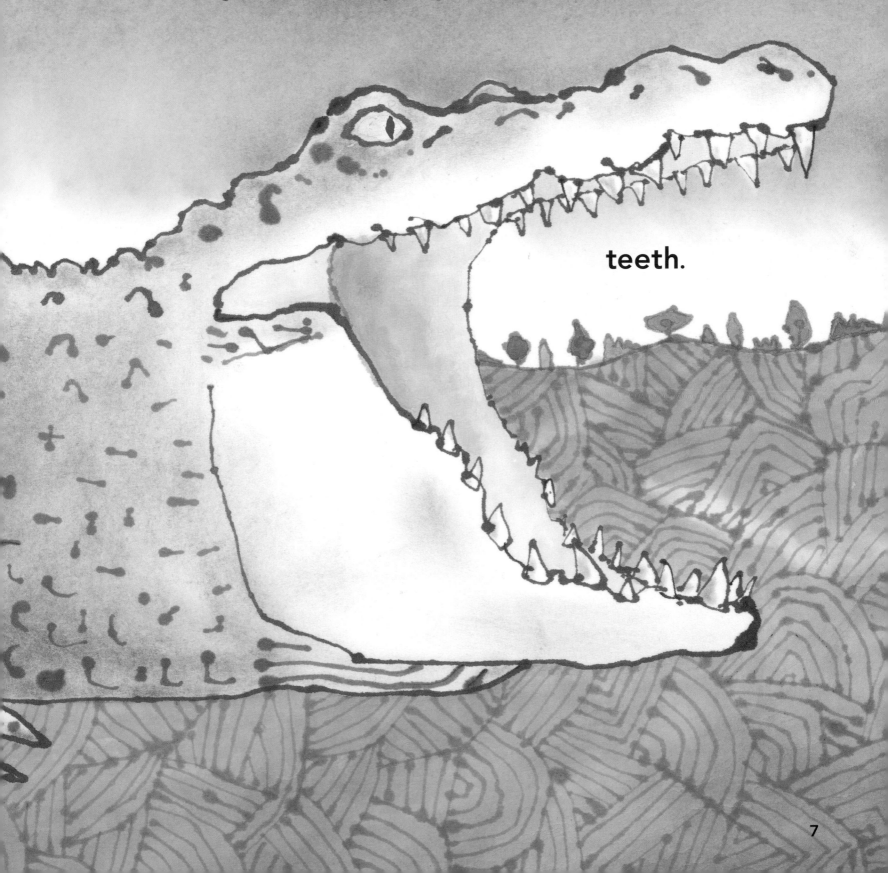

or at least the big ones are. They've got an awful lot of . . .

teeth.

And they're not at all picky about what they eat,
as long as it's got a bit of meat on it.

When it comes to hunting down their dinner,
they're very determined and very cunning.
They know all the places along the sides of the rivers
and lakes where animals come down to drink.

When it's time for a meal, a hungry crocodile will choose one of those places and hide there in the water, just under the surface, with only the top of its head sticking out.

Sooner or later, something passing by,
something with a bit of meat on it,
will decide that it's thirsty and needs a drink.
And then?

Then there'll be a sudden **lunge** and a tremendous

SPLASH.

And then?

Oh, dear.

What happens next is rather gruesome.

In fact it's so gruesome that we should skip the details.

Let's just say there's a lot of **twirling** and **thrashing,**

and then things go a bit quiet.

Afterward, the crocodile won't need to feed again for a while. Instead it spends its time cruising around, checking out places where it might find its next meal, or snoozing on a sandbank.

Crocodiles can go for weeks and weeks without eating. The bigger a crocodile's meal, the longer it can go before the next one.

But there's more to crocodiles than *SPLASH*, snap, twirl, swallow.

You might be surprised to hear that they make very good parents.

Or mothers, we should say. (We'll talk about the fathers later.)

When she's ready to lay some eggs, a mother crocodile gathers

up a huge mound of fallen leaves.

A crocodile usually lays 40 to 60 eggs at one time, but a big one can lay up to 90.

She scoops out a hollow in the top, lays her eggs there, and covers them with leaves. As the leaves rot, they heat up, keeping the eggs nice and warm. The mother can adjust the heat by piling up more leaves or scraping some away.

Inside the eggs, the baby crocodiles are slowly growing. When they're nearly ready to hatch, they start chirping away like tiny birds. That's a signal for the mother to open up the nest.

Then—very carefully—
she picks up the newly
hatched babies in her
enormous jaws and
drops them in the
water nearby.

It takes 80 to 90 days for the eggs to be ready to hatch.

She still doesn't leave the babies on their own, though.

She stands guard over them in the water for months.

You see, lots of things like to eat baby crocodiles: birds such as storks and herons, snakes, big fish, and, worst of all . . .

other crocodiles—sometimes even the baby crocodiles' own fathers!

Despite the mother's best efforts, a lot of the babies
meet an unfortunate end.

But the ones that survive grow . . .

and **grow** . . .

It normally takes about two years for a baby crocodile
to grow to three feet (about one meter) and another eight
years or so for it to reach six feet
(about two meters).

and **grow.**

Until one day, *they're* the ones lurking in the water by that place on the bank, with only their eyes and nose sticking out . . .

waiting for something—or even *somebody*—to come down to drink.

ABOUT CROCODILES

Crocodiles are reptiles, a group that also includes snakes, lizards, and turtles. There are sixteen different kinds altogether, and they can be found in North, Central, and South America, Africa, and the Caribbean. The ones in this book are saltwater crocodiles, the biggest kind—they can be 20 to 23 feet (6 to 7 meters) long. They are found in South and Southeast Asia, Australia, and on some Pacific islands. They sometimes go out to sea but usually live in big rivers and swamps.

Crocodiles' closest relatives are alligators and caimans. There are two kinds of alligators and six kinds of caimans. One species of alligator lives in China, where it's very rare; the other lives in North America. Caimans live in Central and South America. Both alligators and caimans look a lot like crocodiles—it can be quite hard to tell them apart. The easiest way is to look at their teeth when their mouths are shut (though don't get too close!). If you can see teeth sticking up at the side of the mouth, it's a crocodile. If you can't, it's an alligator or a caiman.

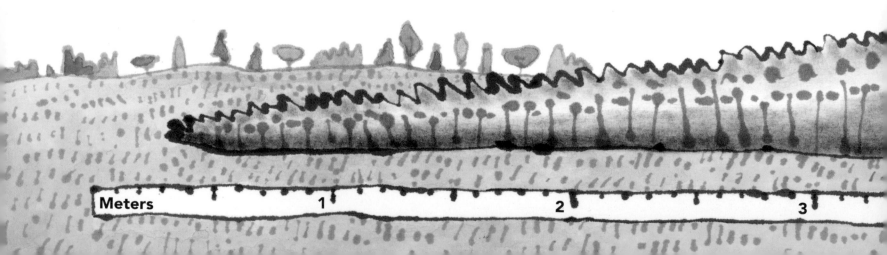

Meters   1   2   3

MORE INFORMATION

A great place to learn more about crocodiles is **www.iucncsg.org**. It's the website of the International Union for Conservation of Nature's Crocodile Specialist Group.

You can also find a lot of useful information at **www.crocodilian.com**.

INDEX

Look up the pages to find out about all these crocodile things. Don't forget to look at both kinds of words: **this kind** and **this kind.**